RIVER FROME
From Source to Sea

RIVER FROME

From Source to Sea

STEVE WALLIS

AMBERLEY

Acknowledgement

I would like to thank Simon Pomeroy for his assistance in photographing the Rainbarrows.

First published 2014

Amberley Publishing
The Hill, Stroud, Gloucestershire, GL5 4EP
www.amberley-books.com

Copyright © Steve Wallis, 2014

The right of Steve Wallis to be identified as the Author of this work has been asserted in accordance with the Copyrights, Designs and Patents Act 1988.

ISBN 978 1 4456 1804 3 (print)
ISBN 978 1 4456 1816 6 (ebook)

British Library Cataloguing in Publication Data. A catalogue record for this book is available from the British Library.

Typesetting by Amberley Publishing.
Printed in Great Britain.

Contents

	Acknowledgement	4
	Map	6
	Introduction	7
1	Through the Chalk Hills	9
2	The Roman Aqueduct	28
3	Past the County Town	39
4	Through Deepest Hardy Country	57
5	The Final Stretch	78
	Bibliography	93
	About the Author	94

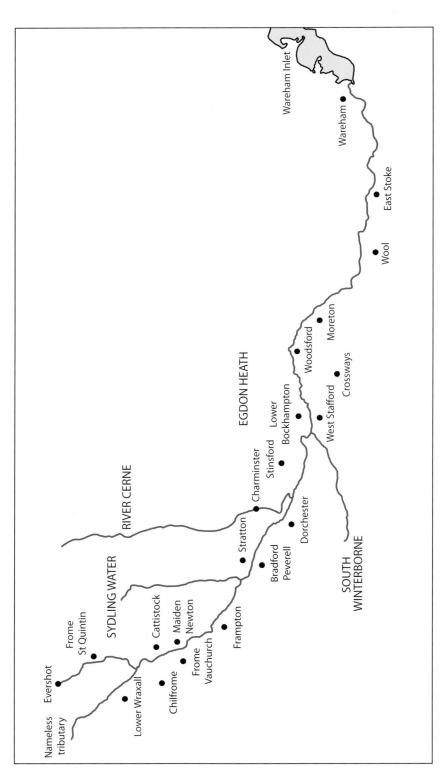

A map showing the route of the River Frome.

Introduction

This book follows the longest river whose course lies entirely within Dorset, describing the Frome and its valley as they pass through central and southern Dorset, together with the historical features associated with them both.

Firstly, a few facts. The Frome is considered to be the largest chalk stream in the South West. It is around 30 miles long and drains a catchment area of 181 square miles, which is approximately one-sixth of Dorset. The river runs roughly south then south-east through the chalklands of south-central Dorset, turning east near Dorchester as it heads through sand and gravel in a broadening valley on its way to Poole Harbour. By the time it passes the county town, its flow has slowed and it generally has a number of channels snaking across the valley floor.

The catchment area is the zone drained by the river itself and also its tributaries, the latter making quite a substantial contribution to the total. Much of this land, especially the upper reaches, lies within the chalk downs of central and southern Dorset. Here, rainwater saturates and then percolates through the porous chalk and the greensand below, stopping at the impervious Gault clay and emerging at the surface through springs. The considerable depth of porous chalk effectively acts as a temporary reservoir that regulates the rate of flow out from the springs, so that chalk streams such as the Frome generally have a more stable rate of flow than rivers based on other geology. This is one of the reasons why they were so conducive to the operation of the water meadow systems that filled much of the valley further downstream, as we shall see.

The name 'Frome' comes from a Celtic word meaning 'fine' or 'fair'. There are several other rivers with this name in southern England, which is testament to the fact that while incoming Saxon settlers gave their own names to new settlements, they tended to keep the river names that were already in use by the existing population.

Thomas Hardy calls it the 'Var or Froom' in his novel *Tess of the d'Urbervilles*, much of which is set at various locations in the valley. 'Froom' is a fairly typical Hardyesque alteration of a real name for use in his fiction, and also reflects the pronunciation of the actual word. There seems to be no precedent for an alternative name of 'Var', though, unless he was somehow thinking of one of the real settlements in the valley – Frome Vauchurch.

Personally, I like the river and its valley for the considerable variation in character they display in such a relatively short distance. It is possible to get a good, general picture of them in quite a short time, say by driving from Evershot through the lanes via Cattistock to Maiden Newton, then on to Dorchester by the A37. Heading east from Dorchester, there is a choice of lanes past places like West Stafford and Moreton before reaching Wool, from where the A352 runs to Wareham along one side of the Frome valley, affording views of the river at its widest and of the Purbeck Hills beyond. Given clear roads, you could probably do this in an hour or so, but if you really want to gain a proper appreciation of the area, I strongly recommend stopping regularly to explore the settlements and indulging in some good walking along the riverside and up onto the flanking hills to appreciate the views. And, of course, cycling is another option for exploration; indeed, the flat countryside of the valley east of Dorchester makes this area very popular with cyclists.

This book is not intended as a gazetteer and covers only a selection of what there is to see, often only that which has caught my interest, so please do explore the valley in all its variety, where you will find much more. I have attempted to describe most of the villages, though when it comes to the two larger towns in the valley, Dorchester and Wareham, these are subjects for whole books in their own right, so I have concentrated on those parts nearer the river.

The Frome has a number of tributaries of varying size and renown. I have not discussed these in any great detail (some, such as the Cerne and South Winterborne, are topics for whole books in themselves), generally describing the section just before the confluence with the Frome.

Particularly lower down, this is a broad valley whose edges are debatable in places, so please indulge my personal interpretation of what is within that valley.

1

Through the Chalk Hills

Our journey along the Frome starts in the attractive village of Evershot, and more precisely just down a back lane, where there is a paved area off to one side with a pair of interpretation boards mounted on a plinth. These tell the story of the village itself and of the river and its geology. While reading them, you might notice a gurgling sound, which is the river emerging from the ground and flowing at the bottom of a channel, though you might not get a clear view because of vegetation. The spot is known as St John's Well, suggesting it once had religious connotations, perhaps due to the gratitude of the villagers, for whom it was the only source of water until connection to the mains in 1979.

It would be a tremendous shame, though, to simply follow the watercourse out of Evershot without looking around the place. It is the second-highest village in Dorset (beaten only by Ashmore up on Cranborne Chase) and has a well-preserved centre with a great many historic buildings. Notable among these is the late medieval parish church at the western end of the village, and houses along the main streets built in terraces that front onto the pavement, giving the impression of a town rather than a village, where you might expect detached cottages.

Not far from the church is The Acorn Inn, which was built in 1664, although much of what we see today dates from the nineteenth century. It was one of the locations used by Judge Jeffreys in his Bloody Assizes that 'punished' the Monmouth rebels in 1685 (a more famous one locally was at Dorchester). It appears under a slightly different name in Thomas Hardy's novel *Tess of the d'Urbervilles*; Tess stops at 'Evershead' for refreshment on her journey from Flintcomb-Ash (over near the head of the Piddle valley) to visit Angel Clare's family at Hardy's Emminster, which is everyone else's Beaminster. Hardy says that Tess does not stop 'at the Sow-and-Acorn, for she avoided inns, but at a cottage by the church'. It is also where Tess encounters the reformed Alec d'Urberville on her return journey.

Summer Lodge lies near the other end of the village, just south of the main street. It dates from 1798, and in 1893 its owner, the Earl of Ilchester, commissioned Thomas Hardy (in his other capacity as an architect) to draw up plans for the addition of the second floor to the building.

The interpretation boards at the river's source.

The Acorn Inn at Evershot.

Evershot parish church with an adjacent cottage that fits the bill for Tess's refreshment stop. Another property on the opposite side of the church also fits the description, and is even called 'Tess Cottage'!

A view of Evershot's main street.

This was only two years after the publication of *Tess of the d'Urbervilles*, which makes me wonder if the Earl wished to show his gratitude to Hardy for publicising the village.

THOMAS HARDY

The Frome valley is inextricably linked with Thomas Hardy (1840–1928) and his novels. Hardy was born and raised in Stinsford, which lies in the valley not far to the east of Dorchester, and spent much of his later life living in Dorchester itself. Like a lot of authors, he set many of his novels in areas he knew, and locations within and around the Frome valley appear regularly as a result. This is particularly the case with the early novels, which are generally set in the area east of Dorchester that he initially just called 'Wessex' after the Anglo-Saxon kingdom of central-southern Britain. Later, perhaps as he ran out of locations, his novels tended to be set in a wider area that covered Dorset and beyond.

The way these locations are used can be fascinating for readers of Hardy's works who are familiar with the areas. Firstly, he changed the names of places. Initially, the change was a major one, so that for instance his birthplace of Stinsford appears as 'Mellstock' and Dorchester as 'Casterbridge'. In later novels, the change is only partial, so that in *Tess of the d'Urbervilles* the name of her birthplace of Marnhull becomes 'Marlott', and the nearby town of Shaftesbury appears as 'Shaston', which is actually a genuine old name for the place.

He then used real locations in and around these towns and villages for action in the novels, making it possible to picture exactly where Hardy imagined things to be happening in his stories. For example, in *Under the Greenwood Tree*, on Christmas Eve the Mellstock village choir clearly follows the path from Lower Bockhampton that runs alongside the Frome before turning up to Stinsford parish church. The characters in *The Mayor of Casterbridge* can be followed around the real Dorchester through the course of the novel, even if Hardy occasionally makes alterations to improve his story.

In other cases, he uses a real location in a more nebulous way to fit his story. A good example is the setting of *The Return of the Native* on Egdon Heath. The latter name is Hardy's own invention for the heathlands on the north side of the Frome valley east of Stinsford, but he makes them much more extensive than they were in his day. He also places the settlement of Mistover and the house at Mistover Knap in the middle of the heath, for which there was no genuine precedent.

The area also features a great deal in Hardy's poetry, with places appearing under the same names as in the novels. This poetry is little-known in comparison with his novels today, but this was his main output from 1895 onwards, when he stopped writing novels following the adverse reaction to *Jude the Obscure*. His poetry was highly regarded in his lifetime because of the way he used the medium to express his innermost emotions.

On leaving Evershot, the river flows south-eastwards between steep-sided hills for a little over a mile, before turning in a more southerly direction. Close to this point, its valley is joined by a railway line, which has just passed through a tunnel bringing it out of the Blackmore Vale, and which will now follow the valley all the way to Dorchester, thereafter heading down to the coast at Weymouth. From a south Dorset perspective, it is usually seen as the line from Bristol, but when constructed in the 1850s by the great engineer Isambard Kingdom Brunel, it left the existing network at Chippenham. Between the tunnel and where we are now lies the hamlet of Holywell, where formerly there was a small station that served Evershot.

Up on the side of the valley near this point is the little junction of roads and tracks called Short Cross, from which point the wonderfully named Sheepwashing Lane descends back down the valley. This is one of those names about whose origin it is entertaining to speculate – my guess is that there was a pond down the lane that was used for this purpose.

Just south of Short Cross, the road that runs down the eastern side of this part of the valley heads into Frome St Quintin, which is the first of several places along

Looking down Sheepwashing Lane from Short Cross.

its course that are named after the river. In such cases, the villages were often differentiated with the addition of the name of the saint to whom the parish church is dedicated, and at first sight this is what it looks like here. However, although the parish church does indeed bear the name of the third-century martyr St Quintin, it was originally named for a family who were lords of the manor here in the Middle Ages and bore the name St Quintin, probably after the place in northern France from which they originated.

The village itself has a number of attractive cottages on one side of the street, with a manor house that was built in 1782 on the other. The parish church itself sits in a lovely and fairly isolated spot on the hillside overlooking the village. Its nave and chancel date from the thirteenth century and its little tower is only slightly younger. In spite of this, one authority has said that the church 'suffered ruthlessly' when it was restored in 1879!

Leaving Frome St Quintin, the twisting road runs past a pair of tall and imposing early twentieth-century cottages. These were built as lodges beside the drive to Chantmarle, a manor house across the valley that dates from the fifteenth century and served as a police training college during the later twentieth century.

Frome St Quintin's parish church.

The Chantmarle lodges.

The road continues through the hamlet of Chalmington, then meets with two others, one of which has come directly from Evershot over the hills on the western side of the valley, the other from the village of Rampisham. Near this junction, the Frome is joined by a tributary, which has been flowing from its source close to the hamlet of Benville in a south-easterly direction through Rampisham and Lower Wraxall. It is not given a name on maps, and I am not sure if it even has a name. Nevertheless, not only is this quite a sizeable watercourse, but I estimate its length from its source to where it joins the Frome is around 6 miles, perhaps 2 miles longer than the Frome at this point. This raises the question as to which of the watercourses in the upper valley is the 'true' Frome, and indeed a case might be made that both of these are tributaries of the River Hooke, which they join a little further downstream. The reality is probably that there are several watercourses in the upper part of the valley that join to form the single river called the Frome, and at some point it was decided that the one flowing from Evershot was the main one, perhaps because its stretch of valley is on the same alignment as the subsequent section.

We are now just outside the large and very picturesque village of Cattistock. Look at the place from one of the many vantage points in the valley and on the surrounding hills, and in all likelihood it will be the tall church tower that will strike you.

A ford across the nameless tributary at Lower Wraxall.

Cattistock from the west side of the valley.

Castle Hill overlooking Cattistock.

However, another historic feature that enhances such views is Castle Hill, which overlooks the village from the north-east. The name 'Castle' comes from an oval enclosure on the hill that is thought to have been Iron Age hillfort; it is not as massive or imposing as some Iron Age hillforts in Dorset, but given its location, perhaps it did not need to be.

Nevertheless, it is the church tower that sticks in the memory after a visit to Cattistock. Like many parish churches, it was extensively restored in the Victorian period. The architect of some of this work in 1857 was Sir George Gilbert Scott, one of the leading lights in the Gothic Revival movement whose works included the Albert Memorial and the Midland Grand Hotel at St Pancras station, both in London. However, it was his son, another George Gilbert Scott, whose restoration work in 1874 was most substantial, including the building of the great tower that we see today.

Chilfrome lies across the valley and to the south-west of Cattistock, about half a mile away as the crow flies. It is a much smaller settlement that clusters around a crossroads and the street that leads to the parish church. The latter dates from the thirteenth century and has a medieval relief of the Virgin Mary above the vestry door. Chilfrome is the second settlement down the valley that bears the name of the river, the other element coming from three noble children who shared the local estate in Saxon times.

Cattistock parish church.

Among the other buildings of note in Cattistock are the village hall, named the Savill Memorial Hall and built in 1926, and the Fox & Hounds Inn.

Pound House in Cattistock dates from the early eighteenth century and has distinctive walls built of chalk blocks.

A pair of locals in Chilfrome.

Above and opposite: There is a good walk between Chilfrome and Maiden Newton along part of the Wessex Ridgeway long-distance path. Much of this route runs beside the Frome. Here we see some autumnal views.

Another mile beyond Chilfrome we reach Maiden Newton, our largest settlement so far, and one of those places that can be difficult to distinguish as a village or a town. It certainly has some of the attributes of a small town; there is a railway station, a road junction, several shops and businesses, and housing developments extending as 'suburbs' for some distance from the old centre. It probably functioned more like a town in the past, particularly as there are no others for the locals to use within a radius of some 8 miles.

From the railway station, the first surviving one that we have encountered, a branch line formerly ran down to Bridport. The line opened in 1857, being extended to the planned holiday resort at West Bay in 1884, and finally closed in 1975.

Down in the centre of the village by a crossroads, there is the surviving shaft of a market cross dating from the fifteenth century. You can just make out the traces of figures carved on it, but they are so weathered that little detail can be seen, and there seems to be no record of who they were.

On the west side of Maiden Newton, past the 200-year-old Maiden Newton Mill, the River Hooke joins the Frome. It has flowed down from the vicinity of the village of Hooke, some 6 miles to the north-west.

Maiden Newton seen from Chammen's Hill on its south-west side.

The market cross.

Off down a lane, we find the parish church, which was renovated recently following a fire. One of its best features is the Norman north doorway.

The Dairy House in Maiden Newton dates from the early eighteenth century, and has walls of knapped flint and banded lias stone.

Maiden Newton Mill.

The confluence of the Hooke and Frome. The Frome is emerging from under the bridge, and because it is also flowing over a weir, it appears that it is flowing into the Hooke, rather than the other way around as convention would have it.

Looking up the Hooke valley from Chammen's Hill.

Historically, the parish of Frome Vauchurch lay just across the Frome from Maiden Newton, and today the two are effectively linked. The 'Vau' element of the place name comes from an Old English word for 'coloured', so the local church was once thought noteworthy for its colour. There are the scattered settlements of Frome Vauchurch and Lower Frome Vauchurch a short distance away, while Higher Frome Vauchurch is really just the part of Maiden Newton that is located on the west side of the Frome.

Opposite: The name of the saint to whom Frome Vauchurch parish church was dedicated has either been forgotten or there never was one. The church has a twelfth-century core and was extensively rebuilt in the seventeenth century.

2

The Roman Aqueduct

After Maiden Newton, the river's course bends from its previous southward direction more towards the south-east as it 'aims' for Dorchester. The valley is broadening, and whereas the bottom used to be generally V-shaped, it is now becoming flatter. The flanking hills are losing height and their sides are becoming less steep.

CONTOUR STRIP LYNCHETS

On the eastern flanks of the valley, just south of Maiden Newton on the west side of Fore Hill, there is a group of contour strip lynchets set in two fields, one diagonally above the other on the hillside. There are other examples of these terrace-like features elsewhere in the valley, but these are probably the most noticeable.

The technical terms need a bit of explanation. Lynchets are features that were formed when sloping ground was ploughed – the soil tended to creep downhill during ploughing, so that after perhaps several centuries this soil would build up against any walls or hedges that ran along, and to a lesser extent down, the hillside. If the area was later taken out of cultivation (for instance, because of loss of soil fertility) and turned over to pasture, then generally walls and hedges would be removed, leaving behind only the banks of soil. These banks are lynchets, and they can be seen on various hillsides around Dorset that have remained under pasture to the present day.

It seems that most areas of lynchets were formed in prehistory, but contour strip lynchets are of more recent origin, and unlike the earlier type they were formed deliberately. They are effectively steps cut into a hillside, forming a series of level terraces, each several yards wide. Most, if not all, are believed to date from the Middle Ages, and their purpose was to allow crops to be grown on land that was otherwise too steep to plough. When you look at them up close, you realise just how much manual labour went into their construction, and what a relatively small area of new cropland was formed on each terrace.

They are testament either to a well-organised and motivated local population, or perhaps to people who had been struggling to survive for a long time and were desperate to bring more land into cultivation.

It is thought that, in many cases, contour strip lynchets did not remain in cultivation for long. A great turning point of the Middle Ages was the Black Death, which had its greatest impact in the middle of the fourteenth century. By wiping out a significant proportion of the population, it not only reduced the number of mouths that needed to be fed, but also increased the price of hired labour considerably; after all, there were now fewer labourers around, so those who were left could demand much higher wages. This led to many landowners taking land out of cultivation and turning it over to pasture, particularly for sheep – one shepherd and his dogs on an area of land would cost the landowner much less than a gang of ploughmen, harvesters and so on.

The contour strip lynchets on the west side of Fore Hill seen from Chammen's Hill. The most obvious lynchets appear as a series of horizontal lines in the field immediately above the houses. These are the bottom right of the two groups; the other to top left is more overgrown and less visible from this angle.

Along this stretch of the valley, and across from the main road from Maiden Newton on the south-west side of the river, there are three scattered settlements. These places (Cruxton, Notton and Throop) are fairly typical of what would have been farming settlements in the Middle Ages, but by the twentieth century had been reduced to little more than single farms. Cruxton has a surviving sixteenth-century farmhouse, while something rather more remarkable was discovered near Throop in the late eighteenth century.

This was an extensive Roman complex, which it has been suggested was of greater importance than the 'ordinary' Roman villa. This complex, like the better known one at Hinton St Mary in the north of the county, had a two-room structure that it is thought could have been a church for the surrounding estate in the late Roman period when the Empire was becoming Christian. This interpretation is strengthened by the discovery of chi-rho monograms on mosaics at both sites (i.e. symbols made up of the first two letters of the word 'Christ' in the Greek alphabet). At Hinton St Mary (the mosaic from which is now in the British Museum), this forms part of what is thought to be the earliest known depiction of Christ in Britain, while at Frampton it was associated with a series of 'dolphin-like fishes' (fish were a symbol of Christ to early Christians).

Shortly beyond Throop we reach Frampton where, like Maiden Newton, there are settlements on either side of the river. The main village is on the north side, while the appropriately named lesser settlement of Southover is on the other. It is not as immediately obvious as other examples, but Frampton also bears the name of the river. Its name means the 'ton' (farm settlement) on the Frome; in this case, the first part has been corrupted to 'Framp' over the centuries.

In Frampton, almost all the buildings are on one side of the road. In 1840, the owners of Frampton Court demolished all those on the south side as they extended their parkland right up the road – the boundary wall stood next to the road until quite recently. The best known of these buildings are the much-photographed row of estate cottages and the parish church. The former are a terrace of four, with a datestone of 1868. The church has fifteenth-century internal features and a tower rebuilt in 1695 by Robert Browne, who later added the north aisle and vestry (this was not a period where much rebuilding of churches took place locally, or even nationally).

Across the river in Southover, there are more historic properties, such as Southover House, a seventeenth-century farmhouse. In a detached location just to the east of the settlement, there is the site of Frampton Court. This was built in the seventeenth century on the site of a Benedictine priory. The main house was demolished in 1935, and the buildings here today are a service wing and a combined stable block and coach house that were both built in the early nineteenth century.

Leaping back to the Roman period, it was long known by antiquarians that an aqueduct ran down the south-west side of the Frome valley to Dorchester. A water channel was known in the area of Southover and was interpreted as part of this aqueduct, leading to the belief that the source of the aqueduct was at Maiden Newton. However, work by the archaeologist Bill Putnam in the 1990s led to the conclusion that the aqueduct's source was a dam in a coombe by Littlewood

Cottages at Frampton.

The estate cottages and parish church at Frampton.

A view across the former parkland of Frampton Court towards the eastern end of Frampton village.

Southover House.

Farm, just down the valley from Southover. The water channel has since been reinterpreted as a water source for the medieval priory at Frampton.

There is an interpretation board beside the site of the aqueduct's dam, which helps in the visualisation of this feature. However, it is in quite a remote spot. One way of getting there is from Southover (there is a pull-in where you can generally park beside the lane that runs from Frampton to Southover, just before the bridge across the river). Take the footpath signposted for 'Muckleford' from Southover – this heads east past Frampton House. After about a third of a mile, there is a junction of paths by Littlewood Farm where you turn right, following the signpost to 'Roman road'. After a short distance, another path turns right and is marked 'Tubbs Hollow'. Follow this path across the field and the interpretation board is just past the gate where the path enters a wood.

The site of the dam that supplied the Roman aqueduct.

THE ROMAN AQUEDUCT

Bill Putnam's work has led to a greater understanding of the workings of the aqueduct. It was first constructed around AD 50 to bring water to the recently established fort at Dorchester. Besides the obvious one of providing water, reasons for its construction may have been to occupy the garrison to prevent restlessness, and to impress the local population with advanced Roman technology.

A first attempt failed because of poor surveying, but the second worked and functioned for about a century, serving first the army garrison then the civil town of Durnovaria. Water was brought from the dam through a channel terraced into the hillsides on the south side of the valley. The channel had a clay base and sides and a cover of timber. Turf was then placed over it to prevent tampering by any but the most determined disgruntled native. It is remarkable to think that the Roman army succeeded in constructing this channel with a gradient of only 1 in 1750, and that to get it from the dam to Dorchester they had to run it around all the coombes on the south face of the valley, leading to a total length of nearly 10 miles to cover only half the distance as the crow flies.

Bill Putnam also showed that around AD 150 it was decided, presumably by the civil authorities in Dorchester, to 'improve' the aqueduct, the channel of which would have required regular repairs. Construction of a bigger channel

A section of the aqueduct terrace not far west of Bradford Peverell.

and terrace began at the Dorchester end, and Putnam suggests that the channel was to be lined with brick or stone. This construction destroyed much of the existing channel, so that when the project was abandoned (perhaps because funding ran out) Dorchester was left without its aqueduct and people had to dig wells or head down to the Frome with a bucket.

It is the terrace of this failed 'improvement' that is seen today, and it seems rather ironic that this is so, considering the skill and ingenuity of the working example.

Today, partly because much of its course has been lost to agriculture and development, but also because where it does survive there is just a terrace on the hillside, the aqueduct is not generally obvious and well known to most people. In effect, you need to know what you are looking for and where it is before you seek it out. This is probably why it does not get as much attention as many other historic features of the Dorchester area. You have to take visitors to see it, but once they 'get' it, and start to think about the engineering involved, they are usually impressed. As Bill Putnam's work has highlighted, the aqueduct is an extraordinary piece of engineering that must have succeeded in overawing the local population as well as bringing water to the Roman garrison.

Not far beyond Frampton, and just before the small roadside settlement of Grimstone, the road from Maiden Newton joins the A37 from Yeovil. Across on the south side of the valley from Grimstone there is another hamlet – Muckleford. The ford after which it was named was replaced in the nineteenth century by a road bridge, and this ford was not as dirty as you might think; 'Muckle' comes from an old personal name, 'Mucela'.

By Grimstone, the Sydling Water flows into the Frome. This is the first of a pair of major tributaries (the other being the Cerne) that flow in a southerly direction through the chalklands into the Frome. In this case, the river is named after a village within its valley, Sydling St Nicholas. 'Sydling' meant 'large ridge' in Old English, presumably a reference to a particular feature in the surrounding hills. There is also the smaller settlement of Up Sydling by the river's source, and the hamlet of Lower Magiston downstream from Sydling St Nicholas.

Half a mile past Grimstone is Stratton. Although we think a 'street' of less importance than a 'road', in Saxon times the word was used to describe Roman roads, and Stratton means 'the settlement on the street', that is, the Roman road that ran from Dorchester to Ilchester in Somerset. This road ran diagonally across

Grazing horses near Muckleford Bridge.

The Sydling Valley near Lower Magiston.

Part of the modern development in Stratton, with the village green that was provided at the same time in the foreground.

what is now the village, and past Grimstone it largely follows the A37 towards Yeovil. Stratton's village street is of later origin, and became part of a turnpike road in the eighteenth century; the village then gained a bypass in the 1960s. Also noteworthy is the extensive development at the western end of the village, construction of which began in the late 1990s. It has now 'weathered' in rather nicely. This development included a new pub, The Saxon Arms.

Bradford Peverell lies a short distance diagonally across the valley. The first part of its name comes from 'broad ford', telling us how the medieval inhabitants crossed the Frome before a bridge was built here, and 'Peverell' comes from a family who were lords of the manor in the Middle Ages. The Roman aqueduct can still be made out looping around a small coombe behind the village. The parish church, whose spire is something of a local landmark, was almost completely rebuilt between 1849 and 1851 by an architect called Decimus Burton.

3

Past the County Town

We are now almost at Dorchester, but first we pass through an area of countryside that is absolutely packed with historic features.

One thing to mention first is that this section of the river is characterised by multiple channels, perhaps partly because of the nature of the terrain through which it flows, and also because of the digging of artificial channels for millstreams and water management purposes.

In this locality, we find most of the best-preserved sections of the Roman aqueduct. A section just south of the back road from Bradford Peverell to Dorchester forms a large V-shape in a field. Look at this from lower down the slope near Whitfield Farm and there is an optical illusion that makes you think the water must have been flowing uphill in this section, but take my word that it's just the angle you are looking from.

The aqueduct then loops around a long and narrow coombe called Fordington Bottom, and nowadays the western section of the Dorchester bypass also runs down it (designed to go through an ancient gap in the aqueduct to avoid damaging this Scheduled Monument). Many people know Fordington as a once separate village that is now within the eastern suburbs of Dorchester, which leads to the question of why a coombe on the other side of the town bears its name. The reason is that until the early nineteenth century, the Borough of Dorchester was limited to the area within the walls of the Roman town, while the parish of Fordington covered an extensive area to the east, south and west. This all changed in the 1830s, when Dorchester's boundaries were extended considerably to absorb Fordington parish.

Over on the north bank there is the confluence with the River Cerne, possibly the best known of the Frome's tributaries. It is some 10 miles long and, like the Sydling Water, it flows past villages that bear its name, of which the most famous is Cerne Abbas, home of the Giant. Its name comes from the same Celtic word that has been anglicised as 'cairn', so presumably it was originally named after a distinctive heap of stones at some now lost location. The last village through which it flows also bears the name, although not so recognisably. This is Charminster, which was

The section of the aqueduct near Whitfield Farm.

first called Cerneminster (i.e. 'the minster church on the Cerne'), and which lies only a few hundred yards north of the confluence. The tower of this particularly picturesque parish church was built in the early sixteenth century by Sir Thomas Trenchard of Wolfeton House, and inside the church there are a number of tombs of the Trenchard family in the south chapel.

Wolfeton House itself lies down a lane on the south-east side of Charminster. The house is in private ownership, but you can follow a footpath part way down the lane. This footpath then bends off to the east across the top of a field, and in that field you can make out the earthworks of the former hamlet of Wolfeton. The house itself dates from the early sixteenth century. A number of alterations and additions, such as a gatehouse, were added in 1862 by an owner named W. H. P. Weston. A datestone of 1534 that was reused in Weston's gatehouse suggests an exact date for the original construction of the house. One of Wolfeton House's outbuildings deserves particular mention. This is the late sixteenth-century Riding House, believed, as the name suggests, to have been used for horsemanship.

Back across the valley, the aqueduct bends round to the east after Fordington Bottom, heading towards the Iron Age hillfort of Poundbury Camp. Below this

The ford on the Cerne at Charminster seen in comparative views, one taken in June (*above*) and the other during flooding in January (*below*).

Charminster parish church.

Earthworks of the deserted settlement at Wolfeton.

section, there is what looks like a large, isolated brick wall. It sits in the valley bottom, and though you might think it is the sole remaining part of some building, it is in a spot that looks too flood-prone for any sensible person to have ever considered inhabiting it. In reality, it was a military firing range, constructed in the nineteenth century for the use of the soldiers stationed at what is now the former Army barracks on the western side of Dorchester town centre on Bridport Road. Much of the layout and buildings of the barracks survive today, and the best-known feature is its former gatehouse, which is now The Keep Military Museum.

The aqueduct skirts around the north and east flanks of Poundbury Camp before disappearing into another coombe, now occupied by industrial buildings, and finally heading into the town. If you ask most people about an Iron Age hillfort at Dorchester, they would answer 'Maiden Castle', after the magnificent example out on the town's south-west side. But in Poundbury Camp, the town has this second, and rather fine, example. Like Maiden Castle, this one also has a complicated history, for its site was also occupied by Bronze Age barrows, one of which can still be seen, and the hillfort was reoccupied around the end of the Roman period. If you approach the hillfort from Dorchester along Poundbury Road, you see it rising up before you

The aqueduct, firing range and railway line seen from Poundbury Camp.

Looking towards the south-west entrance of Poundbury Camp.

to some degree, but you have to go onto the hillfort, especially its northern side, to appreciate the strategic location. It has superb panoramic views across a wide area and a particularly commanding view over the River Frome below.

The railway line that has been accompanying us almost since Evershot disappears into a tunnel that runs diagonally underneath Poundbury Camp from its north to its east side. Brunel's original plan had the line going through a cutting across the hillfort, but objectors who wished to preserve this monument persuaded him to use the more expensive option of a tunnel. The objectors included the well-known Dorset dialect poet William Barnes, whose statue stands outside St Peter's church in the centre of Dorchester. Further south in the town, Brunel had wished to demolish another of the town's prehistoric monuments, the Neolithic henge of Maumbury Rings, as part of his plan to link his new line with the existing one from London, but the same group also persuaded him to relocate the junction and save the Rings.

The industrial estate on the eastern side of Poundbury Camp was built on the site of a First World War camp for German prisoners of war (you can see a memorial to those who died there in Fordington cemetery, which was set up by their comrades after the war). The buildings of Poundbury Industrial Estate are on individual terraces; generally, these are the terraces on which the huts of the prisoner-of-war camp were constructed, the land here otherwise having too much of a slope for building purposes.

In the early 1970s, when the industrial estate was being expanded, extensive cemeteries belonging to Roman Dorchester were excavated on this site by archaeologists. Many other Roman burials have been found at various times in this general area, all the way up to the Bridport Road, as well as at other locations just outside the Roman town. Roman law forebade the burial of people within a town (with the exception of infants), so in common with towns in the Roman world, visitors to Durnovaria in the later Empire would have seen cemeteries extending for some distance outside the town. Those here on the west side seem to have been the most extensive.

We are now reaching the county town, which, together with its environs, is arguably the most historic part of Dorset, not just the Frome valley. Many visitors quite rightly admire its Roman remains and centuries-old buildings that are still in use, but it is arguably the prehistoric features of the locality that make it unique. Some of these are well-known, such as Maumbury Rings and the Iron Age hillfort

Poundbury Industrial Estate, showing the buildings on terraces.

The German memorial in Fordington cemetery, on the opposite side of Dorchester.

of Maiden Castle, but there is much more besides, such as the Neolithic ceremonial monument found in Greyhound Yard, the many barrows (of which more later) and the complexity of the sites previously mentioned. Maiden Castle also had an earlier Neolithic enclosure and a Roman temple, among other features, while Maumbury Rings was a Neolithic henge, a monument for gatherings of some form, a Roman amphitheatre, a fort in the English Civil War, a place of execution and then of public celebrations and performances, continuing in the latter role today.

The Frome runs just to the north of Dorchester, and it is worth following two of its channels here. The first is the more northerly, and runs past two farms that were once hamlets in their own right, like the examples we encountered just after Maiden Newton. These are Frome Whitfield and Coker's Frome, and while no particularly ancient buildings survive at either farm, earthworks of the medieval settlements have been recorded next to each of them. Two bridges that cross this channel are worthy of note, the first being Blue Bridge on the path from Dorchester to Frome Whitfield, which was constructed in 1877 and is, of course, painted blue. The second is Grey's Bridge, which carries the London Road eastward out of Dorchester. It was built in 1740 and bears the maiden name of Mrs Laura Pitt of

Kingston Maurward House, who paid for it as part of other changes to the road system hereabouts.

The other channel, the Millstream, runs much closer to the town. Not far beyond the Poundbury Industrial Estate, it passes Hangman's Cottage. The name of this very attractive property has been taken to mean that the hangman who carried out the sentences from Judge Jeffreys' Bloody Assizes in the 1680s lived here, but this may only be local legend. It seems more likely that at some time it was simply the home of the hangman for the nearby prison.

Beyond Hangman's Cottage is the nearest section of Dorchester's Roman town walls in between the road called Northernhay and the grounds of County Hall. These earth ramparts and ditches, to which a wall was added a short time later, were perhaps built to keep out army mutineers as much as external raiders, and would have been a sign of civic pride as well. After serving a defensive function for over 1,000 years (they were refurbished for the Civil War in the 1640s), they were turned into fashionable boulevards in the latest Continental style in the early 1700s.

Blue Bridge.

The section of the northerly channel just before Grey's Bridge is a good location for admiring the Dorchester skyline.

Grey's Bridge.

Hangman's Cottage.

The Millstream near Hangman's Cottage.

Around the corner from Hangman's Cottage there are a couple of eighteenth-century bridges, and the path that runs beside the Millstream from here to the end of High East Street is an important leisure facility for Dorchester, even having a small nature reserve on its northern side. On the south side, this section of the Millstream first passes Dorchester prison, which was built in the 1790s on the site of what had been a medieval castle, although most of the prison buildings that we see today are Victorian. It then passes the site of a Franciscan friary that was established around the mid-thirteenth century.

No sign of the friary buildings have ever been found, and there is no obvious sign of the Roman town walls hereabouts either. A clue as to why this is may come from Thomas Hardy's *The Mayor of Casterbridge*, wherein the central character Michael Henchard takes a walk from the bridge at the bottom of High East Street that is clearly following this path as far as the site of the friary. Hardy describes Henchard walking beside a cliff, but while today there is what you might call a small cliff beside the prison, there is nothing of the sort between the friary site and the bridge. I suspect this is because Hardy's cliff and much of the area behind

A winter view of the Millstream near Frome Terrace, showing the level ground on the right where Hardy describes a cliff.

The old Lott & Walne foundry.

were dug out ahead of housing development around Frome Terrace in the later Victorian period.

Beyond the bridge at the bottom of High East Street, the Millstream heads into what used to be Fordington parish. It first runs past a 200-year-old former malthouse, then two buildings further along on the old Lott & Walne foundry, built in the early nineteenth century. Heading out of Fordington a few hundred yards further on, it passes the converted and much altered Fordington Mill and a bridge, both of which date from the nineteenth century. From this spot you can look up to the cemetery to see the German memorial mentioned earlier.

WATER MEADOWS

It is on the section of the Frome from just upstream of Dorchester to around Woodsford – a distance of some 7 or 8 miles – where the extensive systems of water meadows are most prominent. Walking across the area you may notice them as shallow channels in the ground surface, generally running in parallel. After flooding they can be very prominent, especially when standing water fills the channels but does not flood the surrounding land. Recent mapping has shown just how extensive these systems were, covering almost all the base of the valley in the area mentioned above.

Simple systems of diverting water across land so that it deposited nutrients had been in existence since the Middle Ages, but the more complicated systems that we see in the Dorchester area were a result of the more scientific approach to farming from the seventeenth and eighteenth centuries onwards. The main idea was to use a steady flow of water across the land in the winter months to maintain the ground at a high enough temperature to prevent frost and so encourage an earlier growth of grass, and also to provide nutrients from the water. This required a chalk river or stream, which was ideal for

Flooded water meadows seen from Poundbury Camp.

carrying nutrients, with a relatively shallow gradient and the steady flow explained earlier.

Water was brought into a piece of land through a series of channels that branched off the river, which gradually decreased in size as they spread out. These channels had names that varied in different parts of the country; the larger were often called 'mains' and the smaller 'carriers', while a system of hatches, sluices and 'stops' were used to regulate the flow through them. The channels were on small banks that kept them just above normal ground level. Once the water reached the smallest of these channels, it flowed over the sides (or 'panes') of the banks to be carried back to the river through other channels with names like 'bottom carrier' and 'tail drain'. The overall effect was of a constant but shallow flow of water across the ground surface, which was often called 'floating', and must have been quite a sight when in operation over a large area.

There was an annual cycle to all this. 'Floating' usually began around Christmas time and continued until there was sufficient grass growth for grazing to be possible – anywhere from late February to the middle of March. Most of the grazing animals were sheep rather than cattle – the latter tended to do a lot of damage to the channels because of their size. These sheep were only kept on the land until early May, because damp pastures in warm weather were a breeding ground for liver fluke. Then the land would be floated again for a week or two to encourage further growth of grass that could then be cropped as hay for winter feed for animals. Cattle would then be brought onto the land for a short period to graze on what was left after the hay cutting (the word for these leftovers was 'aftermath', which nowadays has connotations of what follows a major disaster). During the autumn, all animals would be kept off the land and the channels would undergo removal of silt deposits and repairs of damage caused by the livestock.

The overall effect of the water meadow systems was not just greater productivity of the land. The fact that sheep could be put out to graze earlier in the year than would otherwise be possible meant that far fewer of them had to be slaughtered each autumn since less feed was needed for them over winter.

The men who worked on the water meadows were often called 'drowners', because they literally drowned the land. One local farm of average size would employ four men who worked solely as drowners for several months of the year.

Thomas Hardy describes the work of such drowners, or 'watermen' as he calls them, in *Tess of the d'Urbervilles*. In chapter XXXI, he tells how, when Tess and Angel Clare go for a walk on what he calls the 'meads' one autumn afternoon:

> Men were at work here and there – for it was the season for 'taking up' the meadows, or digging the little waterways clear for the winter irrigation, and mending their banks where trodden down by the cows.

An example of the sluices and hatches that controlled the flow of water on the waterworks by Dorchester.

After developing for a couple of centuries, the water meadows reached their heyday in the nineteenth century, but began to decline in importance around the 1860s. A number of factors caused this, such as the development of artificial fertilisers that superseded the nutrients being deposited by the water flow; increasing costs of agricultural labour; the conversion of many upland areas to arable farming, which meant there were less sheep to use the water meadows; and the formation of the railway network that could transport cattle to markets more quickly and cheaply, encouraging farmers to turn their water meadows over to cattle grazing.

Nevertheless, water meadow systems continued in use in the Dorchester area until after the Second World War, and there was an experimental

A further example of a sluice and hatch near Dorchester.

reintroduction of them just east of Dorchester in the 1950s, which explains why many of the structures that survive in that area are made of concrete!

Although some people in this valley and elsewhere in Dorset (and there is also a well-known example at Harnham close to Salisbury Cathedral) still endeavour to maintain water meadow systems for interest or nature conservation reasons, the fact that they can only do so for a small area is testament to the great engineering skill of those who constructed them, and their successors who kept them operating.

Sheep grazing on water meadow earthworks.

Flooded water meadows between Dorchester and Charminster. Flooding of rivers may disrupt our modern lifestyles and cause us to complain, but such relatively rare events give an indication of what valleys were like for much of the year before the large-scale river management of the past few centuries. Our ancestors accepted such events as normal.

4

Through Deepest Hardy Country

Our subject is now flowing over sand, gravel and clay for the remainder of its journey to Poole Harbour, and this geology gives a broadening valley with a level bottom. In fact, in the section a few miles to the east of Dorchester, you often find yourself in locations where there are no views of distant hills, so that the valley can seem East Anglian in its flatness. Some of the best walking routes in the valley are found here, and the area is popular with people strolling out from Dorchester.

Less than a mile from the eastern fringes of Dorchester we reach the village of Stinsford, which sits on slightly higher ground just north of the water meadows. Stinsford, together with the nearby settlements of Upper and Lower Bockhampton, is the area where Hardy was brought up and figures particularly strongly in his early works.

When Hardy died at Dorchester in 1928, he was a nationally important figure for his poetry as well as his novels, and it was decided that his ashes should be interred in Poets' Corner in Westminster Abbey. There was also a wish to commemorate his strong local connections, and to do so a local vicar suggested that his heart should be interred at Stinsford, which was also the parish church for Upper Bockhampton where Hardy was born. You can still see the tomb Hardy's heart shares with his two wives in Stinsford churchyard, flanked on one side by the tomb of his parents and on the other by that of his brother and two sisters. Other Hardy family graves are close by, as is that of the poet Cecil Day-Lewis.

The path taken by the choir in *Under the Greenwood Tree* leads onward from Stinsford to Lower Bockhampton, where more fine cottages line the street that runs down to the river. Round a corner there is the former village school, now a private residence, which was the prototype for the schoolhouse where Fancy Day taught in that same novel. The street crosses the river over Bockhampton Bridge, which is about 200 years old and which local legend states was built by Hardy's grandfather. This is plausible enough, since both his father and grandfather were local builders.

Behind the path we have just followed lie the grounds of Kingston Maurward, a place with a long history that began as an Elizabethan manor house. A new house

The graves of the Hardy family.

was built on a more prominent site in the eighteenth century in brick, but was refaced in stone after a disparaging remark made by the visiting King George III. The grounds were used by the American army as it prepared to invade France in 1944, and after the war the place became an agricultural college.

We leave the valley bottom now for a short excursion to the lands on the north side of the valley, beyond Hardy's birthplace at Upper Bockhampton. This area is sometimes described as the South Dorset Downs, and today it is covered in plantations and heathland (the latter recently restored for nature conservation reasons in some places). When Hardy was a boy it was almost all heathland, though during his lifetime he commented on the spread of plantations.

One of Hardy's early novels, *The Return of the Native*, is set almost exclusively on this heathland. The central family of the story, the Yeobrights, clearly live at Hardy's own birthplace on the edge of the heath (just as the Dewey family also did in *Under the Greenwood Tree*), and Hardy amalgamates several areas of real upland heath into the single Egdon Heath, which is the real hero of the story. He uses some real locations on the heath in the work, such as Rushy Pond, next to which he places the imaginary house of Mistover Knap, and especially the three Bronze Age barrows known as Rainbarrows, although he has only a single 'Rainbarrow'.

The riverside footpath between Lower Bockhampton and Stinsford.

Bockhampton Bridge.

The view downstream from Bockhampton Bridge.

Restored heathland at Thorncombe Wood Country Park.

Looking east across 'Egdon Heath' from close to the Rainbarrows.

Here the locals build a bonfire on Guy Fawkes Night and look across the Frome valley to many other parish bonfires, and this is also where the heroine Eustacia Vye signals to and meets her lover Damon Wildeve, the proprietor of the Quiet Woman Inn down in the valley. Today you can look down from the Rainbarrows to properties along the Tincleton Road that include Duck Dairy Farm, which was once the Wild Duck Inn, upon which the inn in the novel was based.

RAINBARROWS

The Rainbarrows are the first example of barrows that we have encountered on this trip. Specifically they are Bronze Age round barrows, a term which not only describes their shape but also distinguishes them from the long barrows of the earlier Neolithic period. Barrows are burial mounds, and while long barrows were for communal burials, round barrows were initially constructed for a single individual, which probably tells us something about how society was changing at the time. The first farmers of the Neolithic period probably thought of themselves as communities – they were, after all, groups of people struggling to make a living using their new methods of obtaining food – and buried their dead in the same monument, while by the Bronze Age there was more focus on individuals, particularly the heads of families or clans, who seem to have been those who were buried in round barrows.

The dimensions of round barrows vary considerably, but they can be several yards wide and are often 10 or 12 feet high. Construction of the barrow began with the burial of the dead person in a hole in the ground. Sometimes they were cremated beforehand, and they were often accompanied with symbols of their power, such as weapons and other personal effects, and also food and drink, perhaps for a journey to the afterlife. Much or all of the material for the mound that was then built over the burial came from a circular ditch that was dug around it; although sometimes other material was brought in from small quarries dug close by for the purpose.

Later, other people were buried in the same mound or close by, sometimes in barrows of their own, and it is generally considered that these must have been members of the same family or clan as the original occupant of the barrow; in effect, they were like a family vault in a church in later times.

In some areas, barrows were simply located on waste ground or land of poor agricultural quality. They took up quite a large area, and the early farmers who constructed them could not afford to give up that much of their best arable or pasture land – you can imagine them thinking that, though they wished to honour a dead relative, they could not do so at the expense of their living family.

However, many were clearly placed in the most prominent locations possible. Among the great many barrows that still survive in Dorset, the greatest concentration are along the South Dorset Ridgeway that separates the catchment area of the Frome from the coastlands towards Weymouth and the coast.

The Rainbarrows are another excellent example of this. On their hilltop, they can be seen across a wide area, and indeed they are intervisible with those on the South Dorset Ridgeway. It is probably not taking the evidence too far to think that the descendants of those buried in the three barrows lived and farmed on the fertile lands below in the Frome valley, perhaps occasionally looking up at their ancestors' burial mounds, knowing that one day they too would be buried there, and perhaps feeling a stronger bond to their land as a result.

More barrows are scattered across the heathlands beside the Frome further east, and we shall encounter a few more examples later on.

The most southerly of the Rainbarrows, and probably that which appears in *The Return of the Native*.

Heathlands were generally formed by excessive cultivation of light soils by early farmers, generally during the Bronze Age. Today, their value for nature conservation is well-known, but in the past there was a tendency to dismiss them as wasteland. Hardy's novel helps us to dispel this myth, for many of its characters are accurately portrayed as earning their living from it, as broom-makers, furze- and turf-cutters, and so on.

A Roman road heads up through Thorncombe Wood Country Park not far from Upper Bockhampton and continues out over Duddle Heath. It was part of the network of military communication and control (the implication to the local populace would have been 'we can get to you very quickly if you cause trouble') on a route from London through Old Sarum near Salisbury, Dorchester and on to Exeter. This section, unlike most others, seems not to have been used in post-Roman times. The reason seems to be the steepness of the hillside up which it runs, which is not a problem for a unit of soldiers in fighting condition, but probably was for some peasant's horse dragging a heavily laden cart. Consequently, later routes, including the present A35, tended to go around to the north and south of this hill. The hillside still has views to Dorchester, which leads

The Roman road on Duddle Heath.

to the thought that this section of road, like the aqueduct, also had a function of overawing the native population. A straight road going uphill would have been very visible among what would have been heathland, and once on the top the route has stunning views across the Frome valley. The glint of sunshine on Roman armour moving at speed along the ridge here would have been a sight to behold (and perhaps fear) for the inhabitants of the valley.

We now drop back down into the valley to the village of West Stafford, which lies about half a mile south-east of Lower Bockhampton. The village has a parish church that looks as though it has been miniaturised somewhat, and a central square of sorts outside the Wise Man pub, to one side of which there is a development built in such a traditional style that visitors do not always believe it is less than twenty years old. West Stafford also has two manor houses: a seventeenth-century one on the east side of the village just beyond the modern development, and just outside the west side of the village is Stafford House, which dates from the sixteenth century and sits within the otherwise abandoned settlement of Frome Belet.

By Stafford House, another major tributary flows into the Frome, the South Winterborne. It is so-named to distinguish it from another Dorset river of that

West Stafford parish church.

A view of the South Winterborne, looking east and downstream near Winterborne Herringston, with the earthworks of Winterborne Farringdon (deserted medieval village) on the hillside beyond.

name – Winterborne is a common name for such rivers on the chalk in southern England and indeed a technical term for them as well, for they often only flow in winter when the saturation levels in the chalk are high.

Whereas the Sydling Water and Cerne have two or three settlements bearing their names along their courses, the South Winterborne once had seven. Most of these still exist and in some cases are quite large, such as Winterbourne Abbas and Winterborne St Martin (the latter is better known as Martinstown, and note how some of these villages are spelled with a 'u'). However, some have shrunk and one – Winterborne Farringdon – was totally abandoned several centuries ago, so all that remains above ground is one wall of the old parish church.

The dairy 'Talbothays' plays a pivotal role in the central part of Thomas Hardy's *Tess of the d'Urbervilles*. It was based on Lower Lewell Farm, which lies about three-quarters of a mile east of West Stafford. Parts of the farmhouse date to the seventeenth century, and there is an eighteenth-century dairy house beside it.

The name 'Talbothays' is commemorated in a house called Talbothayes Lodge part way along the road between Lower Lewell Farm and West Stafford, which

Part of Lower Lewell Farm.

Thomas Hardy designed for his brother Henry and sisters Mary and Kate. It was built around 1890, very close to the time when *Tess of the d'Urbervilles* was published. Talbothayes Cottages almost opposite seem to have been built at the same time and may also be Hardy's work.

In *Tess of the d'Urbervilles*, Hardy describes the excellent pasture of this part of the valley, from which milk and butter are sent to the London market. This was a true state of affairs by the 1880s, when the novel was set; the construction of the railway line from Weymouth and Dorchester to London in the 1840s and 1850s enabled produce to reach the latter while still fresh. Indeed, at one point Angel Clare and Tess take milk from Talbothays to a small isolated station for transport. The description of the route they take and the station indicate that Hardy was thinking of Moreton station, which lies some distance from the village of that name – an unimportant matter, since the station mainly served the local farms, not the villagers.

We now head across to the north side of the valley to look at Woodsford Castle, which is often claimed as England's only thatched castle. It was built in 1335 or soon after, and was originally more of a fortified manor house than a true castle. The building was restored in 1850 by the Dorchester architect John Hicks, who employed Thomas Hardy's father, also Thomas, as the builder. Hicks would later take on Hardy junior as an apprentice architect. The property is now owned by the Landmark Trust.

Back over on the south side of the valley, we find the growing village of Crossways. Its name used to sum it up nicely – it was the junction of five routes, with a few cottages around it. Then, in the late 1950s, it was decided that Crossways was to become a dormitory for the expanding Winfrith nuclear power station further along the valley. Although the nuclear site did not expand quite as much as expected, the expansion of Crossways has continued, and now it functions partly as a dormitory for workers in Dorchester, Weymouth and beyond.

A couple of miles north-east of Crossways, and beyond the railway station that bears its name, we find the village of Moreton. If you pass on the main road through the village, you might think it had only a handful of houses, but actually there are a fair few more hidden away round the back. There is a ford and a long, nineteenth-century footbridge on the Frome at a spot that is popular with local swimmers. Beside the road there is a cemetery detached from the churchyard, in which T. E. Lawrence is buried – better known as 'Lawrence of Arabia'. He died in 1935 in a motorbike accident near his home at Clouds Hill, up on the heathland across the river.

The village also bears a strong imprint of the Frampton family, one of whom tried the Tolpuddle Martyrs in the 1830s. Their home, Frampton House, lies just east of the village, and was built between 1742 and 1745 by James Frampton; it has been little changed since, and so is considered a fine example of its period.

The parish church, hidden away off the drive to Frampton House, was built in 1774 by the same James. It is now famous for its stained-glass windows, which depict subjects ranging from aircraft to galaxies, the work of the artist Laurence Whistler in the 1950s. They were put in to replace the originals blown out by a wartime bomb.

On the hill to the south of the village is a 70-foot-tall obelisk, built in the 1780s by Captain John Houlton in memory of James Frampton.

East and south of Moreton, we encounter more barrows, perhaps not as many as on the South Dorset Ridgeway, but a good number. Two lie past Broomhill Bridge on what was once a prominent ridge overlooking the river, and there are more on the heath to the south-west, such as Tadnoll Barrow and groups on the high points of Whitcombe Hill and Blacknoll Hill on the south-west side of the nuclear site.

Opposite the pair of barrows are the grounds of Winfrith nuclear power station, which was decommissioned in the 1990s, while over to the north lies the army base of Bovington Camp, sited here to take advantage of the adjacent heathland, where the rough ground and scrub (together with a lack of farmers needing to be compensated for the loss of crops) make for excellent training for infantry and tank crews. Here there is also the superb Bovington Tank Museum.

Woodsford Castle, as seen from the adjacent road.

Above and overleaf: The ford and footbridge at Moreton.

The cemetery at Moreton where T. E. Lawrence is buried.

Frampton House, seen from the road to the south.

One of the barrows near Broomhill Bridge.

Now onward to Wool, which, like Maiden Newton, is a place that is difficult to categorise as a village or a town. It is growing, so that its 'suburbs' cover quite a large area, and there are shops in the old centre and elsewhere along the main road. There are also a couple of pubs and two filling stations, although there is not a strong feeling of an urban retail centre. On its western side, the spreading settlement has almost engulfed the hamlet of East Burton, where there are more attractive cottages.

Wool is also a place that you could pass through regularly without realising that it actually has a defined centre, and an attractive one at that. In fact, the old part of Wool is hidden away from the main road and railway, explaining why it is easily missed. This area is worth exploring to see the pretty houses, especially along High Street and Spring Street and up Church Lane, where the parish church is not obvious until you get close because of trees and its lack of a large tower.

Out to the north of Wool and just across the Frome lies Woolbridge Manor, which is one place that people are likely to see from the road through Wool. It appears pretty well as itself in *Tess of the d'Urbervilles*; with a slight change of

Cottages in Spring Street, Wool.

name to 'Wellbridge Manor', it is where Angel and Tess spend their honeymoon. The description in the novel of its location in relation to the village, its proximity to the sixteenth-century bridge over the river and how much of a landmark it was to travellers in the valley in Hardy's day all ring true.

Woolbridge Manor, seen from Wool Bridge.

The manor and bridge with the river in winter flood.

This page and overleaf: A selection of summer views of the Frome near Woolbridge Manor.

Winter flooding of the same area. The usual river channel is running diagonally across the picture.

5

The Final Stretch

Heading out from Wool, we notice the valley broadening yet further, and in the distance there are the Purbeck Hills, which form the obvious southern side of the valley. These hills dispel any lingering feelings that the observer is in East Anglia or some other flat landscape.

The land to the south of the river is an area of historic properties. A short distance outside Wool we find Bindon Abbey, which is named after its original location near Bindon Hill, down on the coast near Lulworth. It was founded in 1149, but by 1172 it was discovered that the community there could not sustain themselves on the land around, so the abbey moved to its present site. Like other such institutions, the abbey was dissolved during the Reformation in the late 1530s, and a private house was built on the site soon after. This in turn was burnt down during the Civil War, and the present house and chapel date from the late eighteenth century. A mill nearby may include elements from the original mill that served the abbey.

There is another religious building that moved location a mile or so down the valley at East Stoke. Here the old St Mary's church was abandoned in 1828 because its location close to the river was susceptible to flooding. A new church was built at the same time on the north side of the river beside the main road between Wool and Wareham, though this in turn was converted into three houses in the 1990s.

The old churchyard can still be visited thanks to local efforts to keep it clear of scrub. Inside the churchyard, you can still see a surviving section of the church structure consisting of the south porch and part of the south wall of the nave. The porch includes stone that was robbed from Bindon Abbey after the Dissolution, and you can also just make out the 'imprint' of the building on the ground. The rest of the building was 'quarried' for reuse, presumably much or all going into the 'new' church. It is easy to get suspicious that the abandonment was a ruse to get a nice new church conveniently located by the road, but the stories of flooding may have been linked to a rise in water levels following the construction of water meadow systems in the vicinity. Earthworks of the original and also long-abandoned village of East Stoke can be made out on the south side of the churchyard.

East Stoke's old churchyard from the east entrance. Note the headstones of those buried here before the abandonment and the surviving porch and wall of the church on the right.

East Stoke's new church seen from close to the old one.

A further mile and a half down the valley lies East Holme, a village hidden away among trees that many people probably don't know about. A lane leaves the Wool to Wareham road nearby then crosses a level crossing and a modern bridge, beside which its predecessor, the seventeenth-century East Holme Bridge, still survives. A narrow lane branches off from this one to reach the village, where there is a ford across a tributary of the Frome and also Holme Priory. The latter is named from an establishment of the Cluniac order that was founded here in the early 1100s. Like Bindon Abbey, a house was built on its site soon after the Dissolution, although only parts of this survive, and the present house dates mostly from the eighteenth and nineteenth centuries.

The north slope of the Frome valley between Wool and Wareham is not particularly steep, and much of the land has been and is being quarried for sand and gravel. In this area, a number of Bronze Age barrows were constructed on land that would have been heathland at the time. A significant historic feature here, although one that can be difficult to recognise, is Battery Bank. This extended for several miles roughly along the northern edge of the valley, and was perhaps a boundary between land either under different ownerships or perhaps used for

Holme Priory, viewed from the road to the south.

different purposes. Its most likely date is the Roman period. Evidence suggests that it was never a continuous feature, and perhaps it simply linked natural features, such as clumps of woodland, that could also help to define the boundary.

The town of Wareham effectively sits on a peninsula, formed by the Frome to the south and the River Piddle to the north. Like the Frome and its tributaries, the Sydling Water and Cerne, the Piddle comes from the chalklands of central Dorset, and in its upper reaches it flows roughly parallel with those tributaries.

Wareham's location is thus a strategic one, the rivers affording protection and nearby Poole Harbour allowing transportation of people and goods by sea with relative ease. There is plenty of evidence of settlement hereabouts in prehistoric and Roman times, while inscribed stones dating from *c.* AD 600 to 800 kept in St Mary's church in the town show the presence of people who spoke Celtic and were Christian. Around AD 700, the Saxon saint Aldhelm built a church here.

A major change, and impetus for the development of a town, took place in the late ninth century as a result of the Vikings. During the 870s, the Anglo-Saxon kingdoms of England had almost all been conquered by Vikings who were mainly of Danish origin. The only survivor was Alfred's kingdom of Wessex, but even this was invaded. The Danish army made Wareham its base for a time around AD 875 and then had significant victories over Alfred. Things turned around after Alfred's victory at the Battle of Edington in Wiltshire in AD 878, with the Danes signing a treaty that left them with control of northern and eastern England beyond Watling Street, while the rest belonged to Alfred. The king still had to establish control over this territory, though, as well as protecting it from further Viking raids and invasions, and he planned accordingly. An important part of Alfred's strategy was the establishment of a series of well-located strongholds that could be used for both defensive and offensive purposes. The contemporary name for these was 'burhs' (which comes down to us as 'borough'), and Wareham was one of these. The earth ramparts called Wareham Walls built under Alfred remain one of the most distinctive features of the town, and they have had considerable impact on its layout ever since. Indeed, Thomas Hardy's name for Wareham is 'Anglebury', which was presumably inspired by these rectilinear defences.

The town that developed from the burh did not fill the whole area within the defences until relatively recently; archaeological evidence suggests that much of the area within the walls was under cultivation until the nineteenth century. Nevertheless, the street pattern of Wareham has formed by branching out from the two main roads that run through the town, one going west to east, the other north to south. Thus, it has a grid pattern.

The obvious exception to this is in the south-west corner of the walled area, and is caused by Wareham Castle. This was a Norman construction, and its keep was demolished within a couple of centuries of its building. The site is now occupied by a private house. The castle's main defence on its south side was the River Frome itself, but on the other side were the inner and outer bailey, each of which was surrounded by a sizeable ditch. After the castle's abandonment, the town encroached onto the site of the baileys, but houses could not be built on the soft

Wareham town walls: west side (*above*) and east side (*below*).

infill of the two bailey ditches, so these became streets. Today, Pound Lane follows the inner bailey ditch and Trinity Lane the outer, and both have distinct curves, which seem out of place in the otherwise rectilinear street pattern.

For much of its history, the town clustered around the central crossroads and down towards the bridge over the Frome. Its development suffered a major blow in 1762, when much was lost in a fire that began at the Bulls Head Inn in South Street (and which is commemorated today by a plaque on Lloyds Bank, which now occupies the site).

Despite this, there are some fine historic buildings, among which are the two parish churches. St Mary's down near the Frome is thought to be that founded by Aldhelm. Apparently much of the Saxon church survived intact until a rebuilding in 1840, but only the west ends of the aisles survive from this period. However, a number of medieval additions, such as chapels and the west tower, do remain. The other, in North Street, is dedicated to St Martin and has a nave and chancel of the eleventh century – probably from the earlier part of that century, before the Norman Conquest.

Another building with a religious origin is the Priory Hotel, much of which dates from the early sixteenth century, when it was built as part of a Benedictine Priory.

The following pages show a selection of the historic buildings in Wareham. Above is the mid-eighteenth-century Lloyds Bank, built on the site where the 1762 fire started.

The Manor House of 1712 in South Street, a survivor of the fire

The mid-eighteenth-century Black Bear Hotel, again in South Street.

The Tithe Barn in St John's Square, dating from the sixteenth century.

Wareham Quay.

RIVERSIDE LEISURE AROUND WAREHAM

The section of river from Wareham down to the mouth is the main leisure area of the Frome. The generally calm and relatively broad river full of meanders makes for ideal conditions for a variety of leisure craft. You see lots of canoes and rowing boats near the bridge at Wareham, while further downstream a couple of sailing clubs have their headquarters and a variety of yachts and motor boats are moored along the shoreline. An occasional excursion boat that has come out from Poole navigates this lowest part of the river.

This is also good walking country, and there are several options for walks. One path heads east along the south bank from Wareham Bridge past Priory Meadows to Red Cliffs, where if you don't want to turn back you can carry on to the village of Ridge. Another path joins the north bank just east of St Mary's church and, though overgrown in places when I followed it in summer, it provides a good walk around the Bestwall peninsula. You follow the meanders along the north shore for 2 miles or so to reach an area where you can look across reed beds to the confluence with the Piddle and wider views of the Wareham Inlet and, on the right, the Arne peninsula that juts out into Poole Harbour. There are then two options for getting back into Wareham: either continue on the same path and follow it around on the Piddle side of the peninsula to the town, or retrace your steps a short way and head straight back along a track that leads to Bestwall Road. The relative shortness of the return along the latter route illustrates that although the meandering outward path may have taken you for a good couple of miles, as the crow flies you did not cover anywhere near that distance!

The area to the east of the Wareham fortifications is known as 'Bestwall', a contraction of 'by east wall'. The name is sometimes used for the whole peninsula that projects out between the Piddle and Frome and their confluence. Some of this area became a gravel quarry in the 1990s, and between 1992 and 2005 a group of local volunteers, often working without pay, were led by Lilian Ladle in a major archaeological excavation that worked ahead of the quarrying. Their work identified a tremendous variety of material, including evidence of farming and settlement on the site from the late Stone Age to modern times, as well as Bronze Age burials, pottery production in Roman times, and an encampment from the Civil War.

A short distance along the south bank of the Frome past Wareham bridge is the location where the Furzebrook Railway reached the river. This was built in 1830, and its alternative name of the Pike Brothers' Tramway records those who paid for it. Its intention was to bring ball clay from the quarries in the Furzebrook area 2 miles south to the river for transportation. Ball clay is rich in kaolin and is still valued for specialist ceramic work today. Originally, it was dug out by hand and traded in ball-shaped pieces, hence the name.

The Priory Hotel with St Mary's church behind.

Above and overleaf: Leisure activities on the Frome at Wareham.

This page and overleaf: Views from the path around the peninsula.

The Pike brothers' father had signed a contract with Josiah Wedgwood to supply that great pottery manufacturer with ball clay to improve his products, and the sons proceeded to run the business successfully. The railway had a slight gradient so that gravity did the work of moving the laden trucks from Furzebrook to the river, and horses then hauled them back. The railway lost business when the locomotive-hauled railway reached Furzebrook, but despite this and increasing road transportation, its use continued until 1957. Today you can still follow a trail along much of the route.

Continuing beyond Ridge, the Frome joins the Piddle at the tip of the Bestwall peninsula and together they pour into the Wareham Inlet or Channel. The title of this book indicates that we are following this river to the sea, but look out from the end of the Bestwall peninsula and you do not see open sea. The Wareham Inlet is just one of several bays and inlets that branch off from the great body of water that is Poole Harbour. With an area of about 14 square miles, this is the largest natural harbour in Europe. Today, much of its northern shore is bounded by the town and suburbs of Poole, while on the opposite shore there are broad expanses of heathland and plantations of one of the wilder parts of Purbeck, but like some of the Frome's tributaries and towns, this is a subject worthy of whole books in itself.

The Furzebrook Railway near Ridge.

Looking out across reed beds to the Wareham Inlet from the tip of the peninsula.

Bibliography

An Inventory of Historical Monuments in the County of Dorset, Volume Two – South-East (London: Royal Commission on Historic Monuments, 1970)

Draper, J., *Dorset, The Complete Guide* (Wimborne: The Dovecote Press, 1986)

Farwell, D. E. and T. L. Molleson, *Excavations at Poundbury 1966–80, Volume II: The Cemeteries*, Dorset Natural History & Archaeological Society Monograph Series, 12 (Dorset: Dorset Natural History & Archaeological Society, 1993)

Hardy, Thomas, *Tess of the d'Urbervilles* (Wordsworth, 1993)

Hardy, Thomas, *The Mayor of Casterbridge* (Wordsworth, 1994)

Hardy, Thomas, *The Return of the Native* (Wordsworth, 1995)

Hardy, Thomas, *Under the Greenwood Tree* (Penguin Popular Classics, 1994)

Hutchings, M., *Hardy's River* (Sherborne: The Abbey Press, 1967)

Ladle, L. and A. Woodward, *Excavations at Bestwall Quarry, Wareham 1992–2005, Volume 1: The Prehistoric Landscape*, Dorset Natural History & Archaeological Society Monograph Series, 19 (Dorset: Dorset Natural History & Archaeological Society, 2012)

Ladle, L., *Excavations at Bestwall Quarry, Wareham 1992–2005, Volume 2: The Iron Age and Later Landscape*, Dorset Natural History & Archaeological Society Monograph Series, 20 (Dorset: Dorset Natural History & Archaeological Society, 2012)

Mills, A. D., *Dorset Place-names – Their Origin and Meaning* (Newbury: Countryside Books, 1998)

Newman, J. and N. Pevsner, *The Buildings of England – Dorset* (London: Penguin, 1972)

Penn, K. J., *Historic Towns in Dorset*, Dorset Natural History & Archaeological Society Monograph Series, 1 (Dorset: Dorset Natural History & Archaeological Society Monograph Series, 1980)

Pitt-Rivers, M., *Dorset – A Shell Guide* (3rd ed., London: Faber & Faber Limited, 1968)

Putnam, B., *Roman Dorset* (Stroud: Tempus Publishing, 2007)

Tomalin, C., *Thomas Hardy – The Time-Torn Man* (London: Penguin, 2006)

About the Author

Steve Wallis was born in Halifax in the former West Riding of Yorkshire. He gained a degree in Archaeological Studies from the University of Leicester in 1984 and has since worked mainly in the archaeological field. In 1994, he moved to Dorset to work for Dorset County Council, where he is currently a senior archaeologist. Steve is the author of several titles covering the local history of Dorchester, as well as locations in Somerset and Hampshire. This is his twelfth book relating to Dorset and adjacent counties for Amberley Publishing. He lives in Dorset.

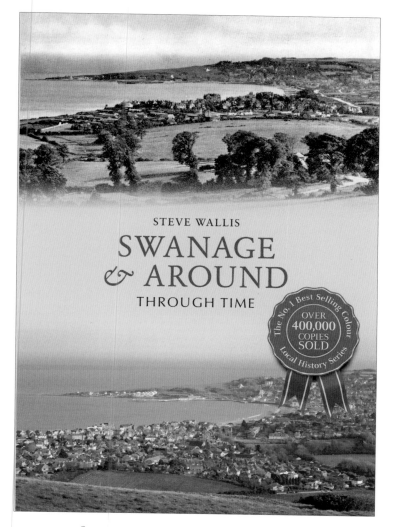

Swanage & Around Through Time

Steve Wallis

This fascinating selection of photographs traces some of the many ways in which Swanage and the surrounding area have changed and developed over the last century.

978 1 4456 1532 5

96 pages, full colour

Available from all good bookshops or order direct from our website www.amberleybooks.com

Literary Portsmouth
Steve Wallis

The city of Portsmouth has a long and fascinating literary history.
In this book, Steve Wallis takes us on a tour of places in and around
Portsmouth that are associated with great literary figures such as
Arthur Conan Doyle, Graham Hurley and H. G. Wells.

978 1 4456 1646 9
96 pages, full colour

Available from all good bookshops or order direct
from our website www.amberleybooks.com